FEAST OF FOOD
2 STATION RD. CAERSWS,
POWYS, SY17 5EQ

Delicatessen and Tea Rooms.
Fresh, Locally produced food,
and produce available
to eat in or take away
OPEN MONDAY TO
SATURDAY 10AM-6PM
TEL; 01686689149

Bragdy Mŵs Piws
Purple Moose Brewery
Porthmadog

the finest ales from north wales
www.purplemoose.co.uk

Bag in Box Limited

Jigsaw Bag-in-Box supply bespoke packaging to hundreds of independent cidermakers throughout England & Wales.
Bag-in-box is the ideal method of dispensing and storing real cider and apple juice because the elimination of air ensures a long, natural shelf life.

www.baginboxonline.co.uk
(01278) 722136

Welsh Mountain Cider

Is made deep in the heart of the Cambrian Mountains. At over 1000 feet above sea level, near the sources of the rivers Severn and Wye, we take the finest fresh juicy apples and squeeze them in Welsh Oak Presses. The sweet juices are left to ferment using their own natural yeasts. This juice is matured in barrels for up to 18 months, producing a unique cider with an uncommonly rich depth of flavour. Our cider contains only 100% apple juice with no added sugars, sulphites or additives. You can find out more about Welsh Mountain Cider, planting apple trees, and making cider by calling 07790071729, or looking at our website: WWW.WELSHMOUNTAINCIDER.COM

Great Oak Glass

Original Stained Glass and Leaded Lights Made to Commission

01686411277
07790071729

www.greatoakglass.co.uk

Welsh Mountain Presses

Cider Presses of all sizes made to commission.
Traditional oak & screw rack & cheese presses.
Steel frame presses.
Basket presses.
Mills & Scratters.
Oak Barrels for sale
All interesting projects considered.

Old presses and paraphenalia bought for cash.

all Bill on 01686411277/07790071729
or email info@greatoakglass.co.uk

If you would like to place an advert in the next edition of this book, call Bill on: 07790071729

HOW TO GROW APPLES & MAKE CIDER

INCLUDING GROW PEARS & MAKE PERRY

by Bill Bleasdale

— WELSH MOUNTAIN BOOKS —
WWW.WELSHMOUNTAINCIDER.CO.UK

PUBLISHED IN WALES BY
WELSH MOUNTAIN BOOKS
PROSPECT ORCHARD
NEWCHAPEL
LLANIDLOES
w.w.w.welshmountaincider.co.uk

Copyright ©2011. William Bleasdale
No part of this book may be transmitted in any form by any means (exept the use of small sections in reviews) without permission in writing from the publisher.

All recommendations in this book are made without guarantee on the part of the author or the publisher. The author and publisher disclaim any liability in connection with the use of this information.

Any resemblance to any person, living or dead, is entirely coincidental.

PRINTED IN LLANIDLOES by AZTEC PRINT
ISBN: 978-0-9566657-1-3

TO MY
BEAUTIFUL WIFE
CHAVA

AND TO THE
VERY KIND & HOSPITABLE
PEOPLE AT EMANDAL
WHERE I WROTE THIS BOOK

"WE WERE NOW GOING THROUGH A COUNTRY FULL OF FINE ENDURING TREES WHERE IT WAS ALWAYS FIVE O'CLOCK IN THE AFTERNOON. IT WAS A SOFT CORNER OF THE WORLD, FREE FROM INQUISITIONS AND DISPUTATIONS AND VERY SOOTHING AND SLEEPENING ON THE MIND."

Flann O'Brien - the Third Policeman

CONTENTS:

① .... introduction
② .... planning an orchard
③ .... apple varieties
⑤ .... rootstocks
⑥ .... planting trees
⑨ .... protecting trees
⑩ .... pruning
⑪ .... pests & diseases
⑬ .... grafting & budding
⑮ .... real cider
⑰ .... perry & pider
⑱ .... the world is your orchard
⑲ .... tumping
⑳ .... making cider
㉑ .... barrels
㉒ .... scratting
㉓ .... presses
㉕ .... notes on equipment
㉗ .... keeping & ageing cider
㉙ .... selling your cider & perry
㉛ .... outroduction.

MYSTERIOUS AND ELUSIVE BLACK CAT OF CAERSWS

WAS IT CONFUCIUS WHO SAID "OF THAT WHICH WE DO NOT KNOW, LET US NOT TALK"? I DON'T KNOW

THERE'S ALSO LOTS OF THINGS I DON'T KNOW ABOUT APPLES, PEARS, CIDER AND PERRY. RATHER THAN COVER THIS BY "RESEARCH", I HAVE PRETTY MUCH ONLY PUT IN STUFF BASED ON MY OWN EXPERIENCES, THE EXPERIENCES OF PEOPLE I KNOW AND TRUST, AND STUFF I'VE MADE UP FOR ARTISTIC EFFECT.

ONCE YOU START TO GROW APPLES & MAKE CIDER YOU WILL FIND THE BEST WAYS OF DOING THINGS FOR YOURSELF.

Planning an Orchard (2)

How many apples, pears & other fruits do you require? Do you want fruit for eating, cooking and making jams & preserves, juice, cider & perry? On a site, planted on standard rootstocks, one can expect 10 tons+ per acre after approx 14 years from planting. 1 ton of apples yields around 120 imperial gallons of juice or cider.

I have planted my main 2 acre orchard on M25 (full vigour) root stock. I like big trees — they have the vigour to throw off disease, they look great and offer habitats for beasties. I don't mind about the size, because I'll be mostly harvesting by
— shaking the trees over tarpaulins —

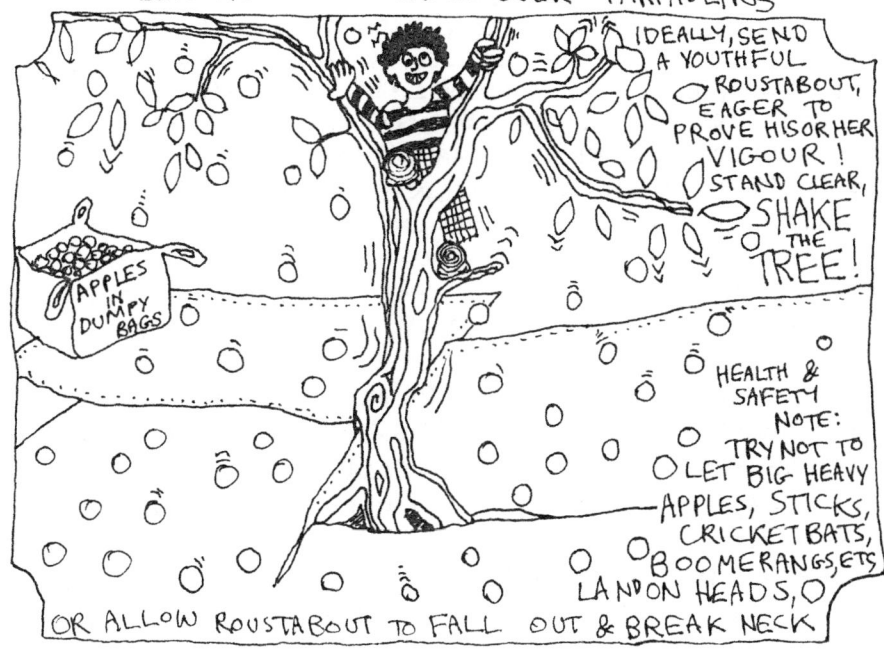

Ideally, send a youthful roustabout, eager to prove his or her vigour! Stand clear, SHAKE THE TREE!

APPLES IN DUMPY BAGS

Health & Safety note: try not to let big heavy apples, sticks, cricket bats, boomerangs, etc land on heads, or allow roustabout to fall out & break neck.

Apple Varieties

I BELIEVE IT WAS WILLIAM RUSHTON WHO SAID "HAVE NOTHING IN YOUR HOUSE THAT YOU DO NOT KNOW TO BE USEFUL OR BELIEVE TO BE BEAUTIFUL."

TO EXTEND THAT WISE ADVICE TO THE GARDEN WOULD BE TO BECKON IN THE NOBLE PYRUS & THE MIGHTY MALUS. AS A MATTER OF COURSE. HERE'S SOME OF THE APPLE VARIETIES I'M GROWING — I INTEND TO CONSTRUCT AN ELABORATE POMONA AT SOME POINT IN THE FUTURE......

ACKLAM RUSSET • ADAM'S PERMAIN • ARDLAM RUSSET • ASHMEAD'S KERNEL BAKER'S DELICIOUS • BALL'S BITTERSWEET • BANANA PIPPIN • BANWELL SOURING BEAUTY OF BATH • BEAUTY OF CORNWALL • BELL APPLE • BELLE NORMAN BEN'S RED • BICKINGTON GREY • BILL NORMAN • BILLY DOWN PIPPIN BLACK DABINETT • BLOODY PLOUGHMAN • BLOODY TURK • BLUE SWEET BRAEBURN • BRAMLEY • BRAMLEY ORIGINAL • BREAKWELL'S SEEDLING BROOM APPLE • BROWN SNOUT • BROWN'S APPLE • BROXWOOD FOXWHELP BUTTERY D'OR • CAERSWS CIDER • CAMBUSNETHAN • CAMELOT • CAP! BROAD CASTLETON CANAL • CHAXHILL RED • CISSY • COLEMAN'S SEEDLING COURT DE WYKE • CRIMSON GRAVENSTEIN • CRIMSON VICTORIA • DABINETT DEVON RED • DISCOVERY • DOOZER'S PROLIFIC • DUFFLIN • DUNKERTON LATE SWEET EARLY JULYAN • EASTER ORANGE • EDWARD VII • EGREMONT RUSSET ELLIS BITTER • ELLISON'S ORANGE • FAIR MAID OF DEVON • FILLBARREL FREDERICK • GEORGE CAVE • GILFACH • GLANSEVIN • GLOSTER • GOLDEN BALL GOLDEN SPIRE • GRENADIER • HALSTOW NATURAL • HARRY MASTERS JERSEY HEREFORD REDSTREAK • HEREFORD RUSSET • HERRINGS PIPPIN • HUGHES CARIAD • IAN COSTARD • IMPROVED REDSTREAK • IRENE'S FAVOURITE IRISH PEACH • JENKINS № 1 • JOE'S SMALL RED BITTERS • JOHNNY ANDREWS

④

JONAGOLD · KESWICK CODLIN · KILL BOY · KILLERTON SHARP · KILLERTON SWEET
KING OF THE PIPPINS · KINGSTON BITTER · KINGSTON BLACK
LADY SUDELY · LADY'S FINGER · LANGWORTHY · LAXTON'S FORTUNE
LINK WONDER · LLANACHAERON BEAUTY · LLANACHAERON PEACH
LLANDINAM PERMAIN · LLIDY STRIPY CRAB · LONGSTEM · LORD BURGHLEY
LORD DERBY · LORD GROSVENOR · LORD LAMBOURNE ·
LORD OF THE ISLES · MACHEN · MERTON BEAUTY · MICHELIN
MONMOUTH GREEN · MORGAN SWEET · MRS PHILLIMORE · NONPAREIL
NORFOLK BEAUTY · NORTHWOOD · ORLEANS REINETTE · PAIGNTON MARIGOLD
PEAR APPLE · PEASGOOD NONSUCH · PENGALED · PERTHEYRE
PHELP'S FAVOURITE · PIG ADERYN · PIG YR WYDD · PIG'S NOSE
PIG'S SNOUT · POLLY · PONSFORD · PRENGLAS · RAGLAN RED
RED DEVIL · RED FALSTAFF · RED STYRE · RED WINDSOR
REVEREND WILKES · RIBSTON PIPPIN · ROYAL JUBILEE
ROYAL RUSSET · ROYAL SOMERSET · ROYAL WILDING
SAM'S CRAB · SCOTCH BRIDGET · SCRUMPTIOUS · SEVERN BANK
SHEEP'S NOSE · SLACK MA GIRDLE · SOMERSET REDSTREAK
SPOTTED DICK · ST CECILIA · STANTWAY KERNEL · STIRLING CASTLE
STOKE RED · SUNTAN · SWEET ALFORD · SWEET CLEAVE
SWEET COPPIN · TALE SWEET · TAN HARVEY · TAUNTON FAIR MAID
TEN COMMANDMENTS · TOM PUTT · TOWN FARM No. 59
TREGONNA KING · TREMLETTS · TWLLDYN GWYDD · TYLER'S KERNEL
VALLIS APPLE · WELSH DRUID · WERN · WHITE JERSEY
WHITE NORMAN · WHITE WICK STYRE · WHITPOT SWEET
WILLIAM CRUMP · WINTER QUOINING · WINTER STUBBARD
WORCESTER PERMAIN · WYKEN PIPPIN · YELLOW STYRE

⑤ APPLE ROOTSTOCKS....

~ APPROX HEIGHT AFTER 10 YEARS...

M25   M27   M26

PEAR ROOTSTOCKS.....

QUINCE "C"   QUINCE "A"   SEEDLING PEAR

I LIVE IN A VERY WINDSWEPT AND HIGH PLACE, SO I HAVE PLANTED MY MAIN ORCHARD ON M25 ROOTSTOCK. THIS IS VERY VIGOROUS AND WILL PRODUCE A TREE OF AROUND 20' EVENTUALLY. BECAUSE OF ITS VIGOUR, IT THROWS OFF DISEASES AND PESTS MORE EASILY THAN A DWARFING ROOTSTOCK. M26 IS A SEMI DWARFING ROOTSTOCK - OFTEN USED ON MANY A GARDEN APPLE TREE. M27 IS VERY DWARFING, AND A TREE WILL RARELY EXCEED 7FT, MAKING IT A GOOD CHOICE FOR CONTAINERS AND WALK-OVER ESPALIERS

THERE ARE OBVIOUS ADVANTAGES IN A SMALLER TREE - EASIER TO PRUNE, THIN & HARVEST FRUIT. I AM NOT REALLY GROWING TABLE FRUIT, AND AM HAPPY TO HARVEST FOR JUICE AND CIDER BY SHAKING THE TREE, HAVING FIRST PUT A TARPAULIN UNDERNEATH IT.

THERE ARE MANY OTHER ROOTSTOCKS AVAILABLE AS WELL AS THOSE LISTED ABOVE - THESE ARE JUST ONES THAT I USE AND AM FAMILIAR WITH

NIDDER CYMRAEG

⑥

A ROUGH IDEA OF HOW MANY TREES TO PLANT PER ACRE.

M25 .... 60 TREES PER ACRE
M26 as half standards .... 115 TREES PER ACRE

MY M25 TREES ARE PLANTED 15 FT. APART IN ROWS 25 FT. APART. THERE MAY BE SOUND ARGUMENTS FOR PLANTING MORE TREES AND THINNING IN A FEW YEARS, BY WHICH TIME THE THINNED TREES WILL HAVE MORE THAN PAID FOR THEMSELVES — AND YOU'LL GET A BIT OF FREE FIREWOOD!

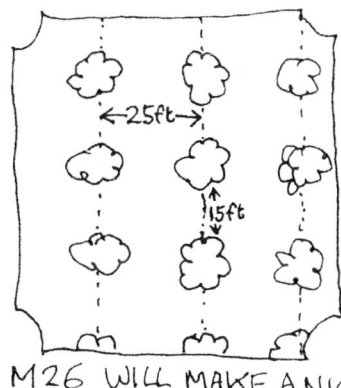

WITH M27 ROOTSTOCKS, YOU COULD EVEN HAVE A MOVABLE ORCHARD IN HALF BARRELS!

BECAUSE I LIVE IN A HIGH, WINDY PLACE, I HAVE PLANTED M26 (RATHER THAN THE LESS VIGOROUS M27) STOCKS 6FT APART AND AM MAKING AN ESPALIERED APPLE MAZE CONTAINING OVER 200 VARIETIES.

STANDARD: 6' TRUNK
½ STANDARD: 4' TRUNK

M26 WILL MAKE A NICE STANDARD OR HALF STANDARD.

EARLIES, MIDS & LATES

AND SO IT GOES...... PEAR & APPLE TREES FLOWER AND FRUIT OVER A PERIOD OF SEVERAL WEEKS, DEPENDING ON THE VARIETY.

SIMILARLY, THE TREES' FRUIT RIPENS AT DIFFERENT TIMES ACCORDING TO VARIETY. WE HAVE EARLY, MID & LATE FLOWERERS & EARLY, MID & LATE FRUITERS. LATE FLOWERERS ARE GENERALLY SAFEST IN FROST POCKETS.

HEDGING, WINDBREAKS, FROST.

WILLOW STICKS ARE SELF ROOTING & WILL THRIVE IF STUCK IN THE GROUND (THROUGH MULCH MAT SO THEY DON'T HAVE TO COMPETE WITH GRASS & TURF). HEDGING PLANTS ARE CHEAP, AND FRUIT TREES WILL DEFINITELY BENEFIT FROM THE PROTECTION. GOOD TO LEAVE GAPS IN DOWNHILL SIDE HEDGES TO LET COLD AIR RUN THROUGH IN WINTER

⑦ PLANTING APPLE TREES

BARE ROOT

IF YOU ARE PLANTING BARE ROOT TREES, YOU WILL NEED TO DO THIS IN THE DORMANT SEASON (AFTER AUTUMN, BEFORE SPRING...) CONTAINER GROWN TREES CAN BE PLANTED OUT AT ANY TIME OF YEAR

POT GROWN

PREPARING GROUND

DON'T! DIGGING A LUXURY TREE PIT FOR YOUR TREE IS: (a) HARD WORK (b) DESTROYS 20 THOUSAND YEARS OF SOIL STRUCTURE. (c) CREATES A BACKFILLED POND FOR TREE ROOTS TO DWINE* IN. FOR BARE ROOT TREES, EMPLOY A "FORESTRY NOTCH" SYSTEM, CUTTING EXTRA SLITS FOR ANY LARGE ROOTS. POTTED PLANTS CAN BE PLANTED IN A HOLE THE SIZE OF THE POT, TAKING CARE TO TAMP EARTH FIRMLY INTO ANY GAPS. STAKING IS UNNECESSARY FOR TREES WITH VIGOROUS ROOTSTOCKS. DON'T FORGET TO MAKE A MAP OF WHICH TREE IS WHERE — LABELS CAN FALL OFF OR FADE!

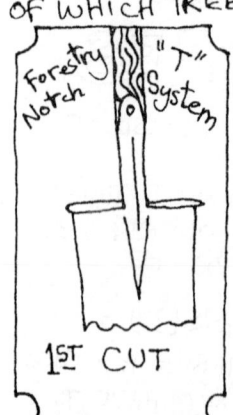
Forestry Notch "T" System — 1ST CUT

2ND CUT / 1ST CUT

PLANT WITH GRAFT AT LEAST 2" PROUD OF EARTH — FOLD BACK FLAPS / PLANT TREE IN HOLE

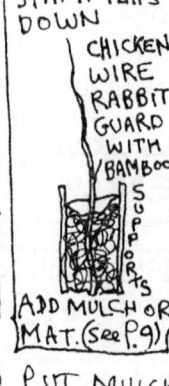
STAMP FLAPS DOWN — CHICKEN WIRE RABBIT GUARD WITH BAMBOO SUPPORTS — ADD MULCH OR MAT. (See p.9)

WHEN PLANTED, PROTECT WITH A TREE GUARD AND PUT MULCH MAT / PEA GRAVEL † MULCH 3 FEET AROUND BASE.

*See "Fungus The Bogeyman" by Raymond Briggs, Hamish Hamilton, 1977

(8) SOME MORE NOTES ON TREE PLANTING

Most apple trees are fairly tolerant of wet soil for a couple of months in the winter, but they will not thrive in permanently waterlogged soil. Signs such as sedge growing, or for that matter, extreme & continual wetness are pointers to poor performance. If you are lumbered with a wet site, you could try mounding up earth 3 or 4 ft. and planting the trees atop of those humps.

As mentioned, I don't advocate the use of big holes. Forestry notch system suffices for small trees.

(It's nearly always better to plant small)

The presence of sedge suggests that this is not an ideal spot for trees.

If there are a couple of long, straggley roots, dig slits to accommodate them.

In my experience, small 1 yr. old trees establish and grow quicker than bigger older ones — they cost less, too, and are easier to plant.

I wouldn't advocate heavy feeding of trees growing in the ground, as this will encourage vigorous, sappy growth. We want wiry resilient growth. I top dress all my trees in year 2 with fish, blood & bone, and occasionally dust with a bit of wood ash for potash.

It is good to keep the grass down in a 6ft radius around the base of your trees.

Protection of Trees (9)

MULCH MAT
TREE POKES THROUGH SLIT IN MIDDLE OF 3' SQ. MAT
CANE & CHICKEN WIRE RABBIT GUARD
WOVEN/FELTED POROUS LANDSCAPE FABRIC
ROUND EDGE SPADED IN

ADVANTAGES – CAN BE LAID STRAIGHT ON TO TURF – NO NEED TO STRIP GRASS. EFFECTIVELY GIVES ROOTS FREE RUN.
PROBLEMS – VOLES LIKE HANGING OUT, MAKING TUNNELS & NIBBLING TREES UNDER IT. GOOD TO ALLOW GRASS TO GROW UP IN 3" SLIT AROUND TREE BASE – KEEPS NIBBLERS AWAY FROM BARK

AS SOON AS YOU HAVE PLANTED YOUR TREES, YOU SHOULD PUT SOME KIND OF GUARD ON – RABBITS CAN STRIP AN IMPRESSIVE NUMBER OF YOUNG TREES IN A NIGHT. IF YOU HAVE DEER OR LIVESTOCK, YOU WILL NEED MUCH MORE SUBSTANTIAL GUARDS. I FAVOUR A GUARD WHICH ALLOWS FREE MOVEMENT OF AIR. A CHICKEN WIRE CYLINDER HELD IN PLACE WITH CANES. **FOAMY BUFFERS** OF GREY SPONGE PIPE LAGGING AROUND THE TOP OF THE CAGE WILL STOP BARK DAMAGE

"ROUNDUP"
CAN APPARENTLY BE SPRAYED AROUND TREE BASE TO KILL GRASS & TING. QUITE POPLAR, BUT WHAT'S THAT JETHRO? JETHRO BE SAYIN' 'TIS SOMETHIN AKIN TO WITCHCRAFT, BEST LEFT ALONE, BUTT. FAIR PLAY!

3" PEA GRAVEL
POSSIBLY THE BEST, THO I'VE NEVER HAD THE SPARE CASH TO TRY IT. IT WOULD LOOK TIDY, AND WOULD BE VOLE TUNNEL PROOF!

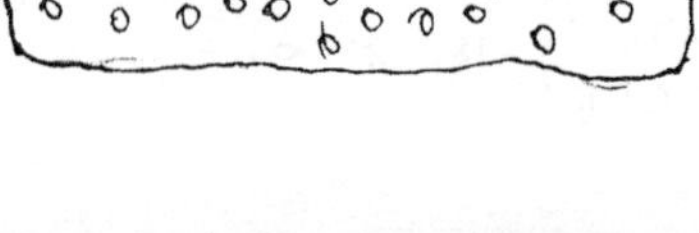

PROTECTING TREES FROM LIVESTOCK

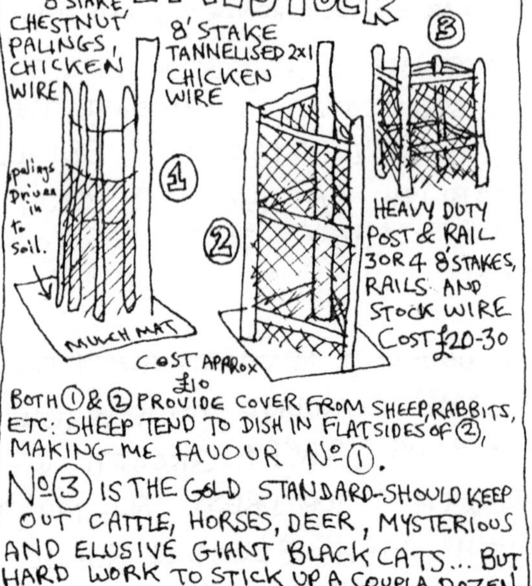

① 8' STAKE CHESTNUT PALINGS, CHICKEN WIRE
palings driven in to soil.
MULCH MAT

② 8' STAKE TANNELISED 2x1 CHICKEN WIRE

③ HEAVY DUTY POST & RAIL 3 OR 4 8' STAKES, RAILS AND STOCK WIRE COST £20-30

COST APPROX £10

BOTH ① & ② PROVIDE COVER FROM SHEEP, RABBITS, ETC. SHEEP TEND TO DISH IN FLAT SIDES OF ②, MAKING ME FAVOUR №①.

№③ IS THE GOLD STANDARD – SHOULD KEEP OUT CATTLE, HORSES, DEER, MYSTERIOUS AND ELUSIVE GIANT BLACK CATS... BUT HARD WORK TO STICK UP A COUPLA DOZEN IN A DAY!

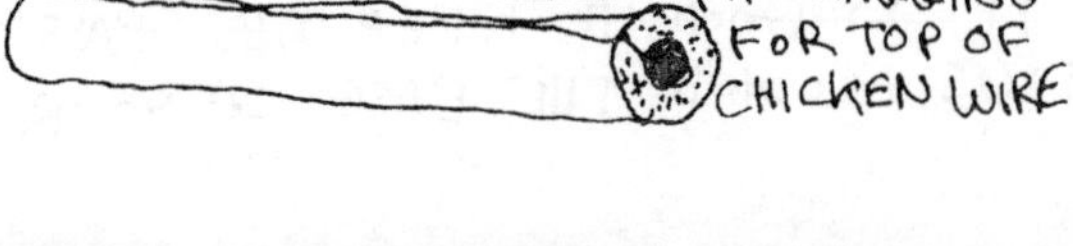

PIPE LAGGING FOR TOP OF CHICKEN WIRE

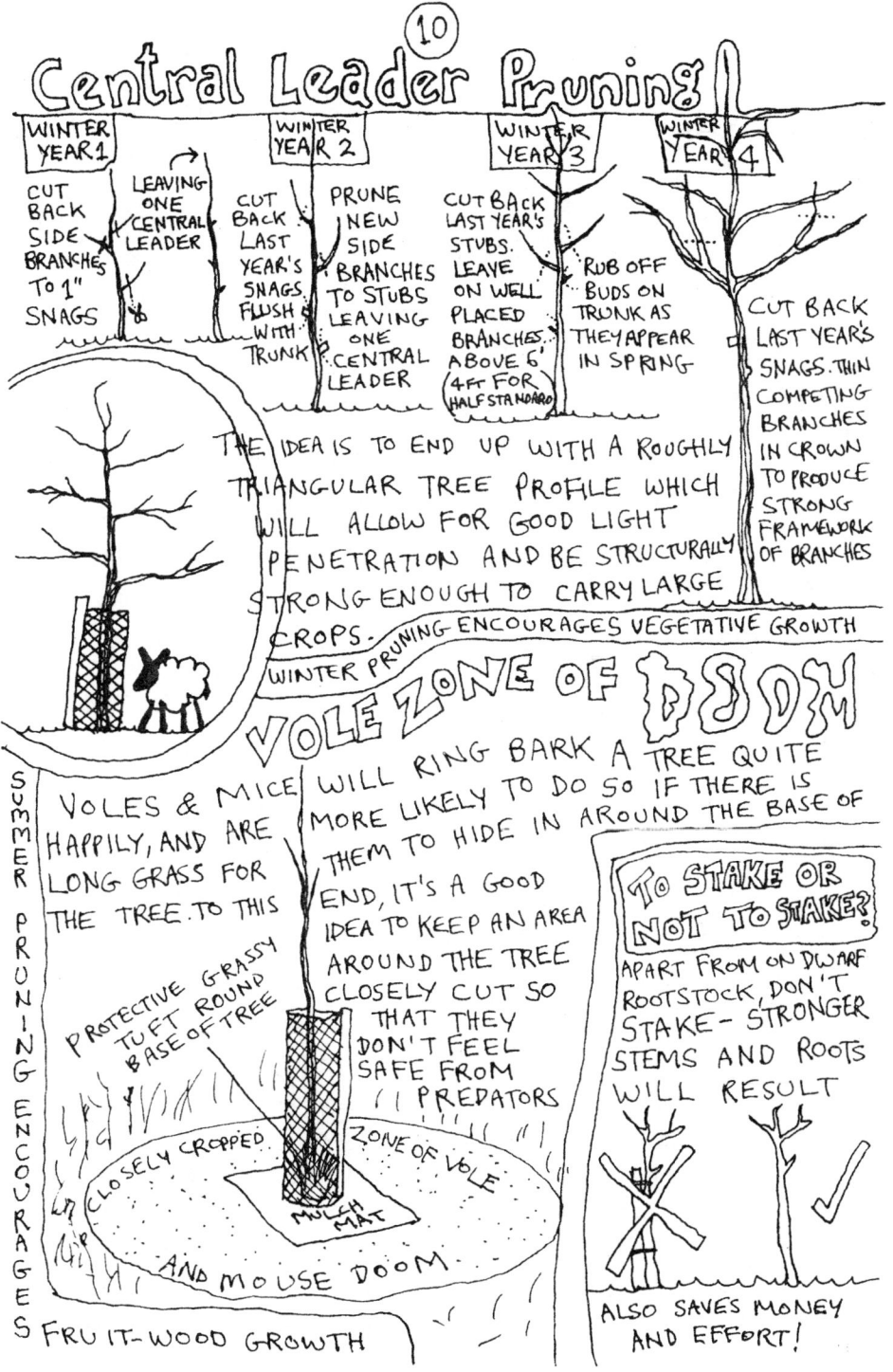

PESTS 'N' DISEASES (11)

Canker

"SCABBY SCARRY BLACK DIE BACK. MOISTY SPORE BORNE DESPOILER OF APPLE STICK AND BRANCHEY" (RABBI BOINS 1809)

CAN BE NASTY. SOME VARIETIES MORE SUSCEPTIBLE THAN OTHERS. PRUNE OUT & BURN AFFECTED PARTS, ME DEARS. REALLY GRIM CASES; DIG OUT & BURN TREE. WHEN PRUNING, DISINFECT SECATEURS BETWEEN EACH TREE AS A PRECAUTION (METHYLATED SPIRITS IS FINE....). HAVING SAID THAT, I KNOW A CHAP HEREFORD WAY WHOSE BUSH TREES ARE COVERED IN CANKER, AND WHO TOOK 20 TONS OF APPLES PER ACRE OFF THEM LAST YEAR! SO MAYBE JUST LEAVE IT!! OR NOT!! GOT THAT? GOOD.

Fireblight

MORE COMMON AND DAMAGING ON PEARS THAN ON APPLES. MAKES BLOSSOMS BLACKEN AND CAN CAUSE BRANCH TIPS AND EVENTUALLY THE WHOLE TREE TO BLACKEN & DIE AS IF IT HAD BEEN ON FIRE

CUT OFF AFFECTED AREAS 6" INTO HEALTHY WOOD & BURN.

(12)

WELL, REALLY! YOU WANT MORE? BEST NOT TO KNOW 'TIL IT HAPPENS. BIT LIKE HAVING A COPY OF THE "HOME DOCTOR" ON THE SHELF.

GO ON, YOU CAN HAVE ANOTHER COUPLE, BUT THAT'S YER LOT — YOU CAN LOOK UP THE OTHERS FOR YOURSELVES.

WOOLY APHID

not actual size, shape or resemblance of wooly aphid

"KNIT ONE, PEARL ONE."

BE APHRAID, BE VERY.....

"OH SAUCY SAP SUCKER OF THE WOOLY VARIETY, YER FELTY FRATERNITY DOTH MAKE THE LEAF TO CURL AND THE STRAIGHT LEAD TO KINKIFY......"

BILL COBBET, WOODLAND WANDERS 1721

DON'T DO MUCH HARM IN MY RECKONIN. STILL, WHAT DO I KNOW? SOMETIMES CAUSE NEW GROWTH TO KINK & SPIRAL AS WELL AS LEAFY CURL. YOU CAN ATTACK THEM ON A SMALL TREE WITH SPIRIT OR SUMFINK. I WOULDN'T BOTHER, THE TREE'LL SORT ITSELF OUT.

SCAB 'n' CODLING MOFF, etc.

ARE YOU STILL HERE?

"GIVE ME SCABS ON MY APPLES, BUT LEAVE ME THE BIRDS & THE BEES PLEASE!!"

JUST CODLING ALONG, JOHN

Joni Mitchell 1970

WE DON'T REALLY CARE ABOUT ANY OF THIS, DO WE? COZ WE'RE GOING TO MAKE JUICE AND CIDER AND TING OUT OF OUR FRUIT. YOU CAN LOOK IT UP SOMEWHERE ELSE IF YOU'RE INTERESTED. SEEM TO RECALL SOME TALE OF AN ORGANIC ORCHARDIST TURNING TURQUOISE FROM TOO MUCH COPPER SPRAY! AARGH!! CLASS DISMISS!

GRAFTING
VERY SHARP KNIFE

FOR OUR PURPOSES, GRAFTING IS THE JOI TOGETHER OF A ROO WITH CERTAIN CHARACT AND A SCION (A PIECE VIGOUROUS WOOD FROM VARIETY YOU WISH PROPOGATE).

GRAFTING TAPE

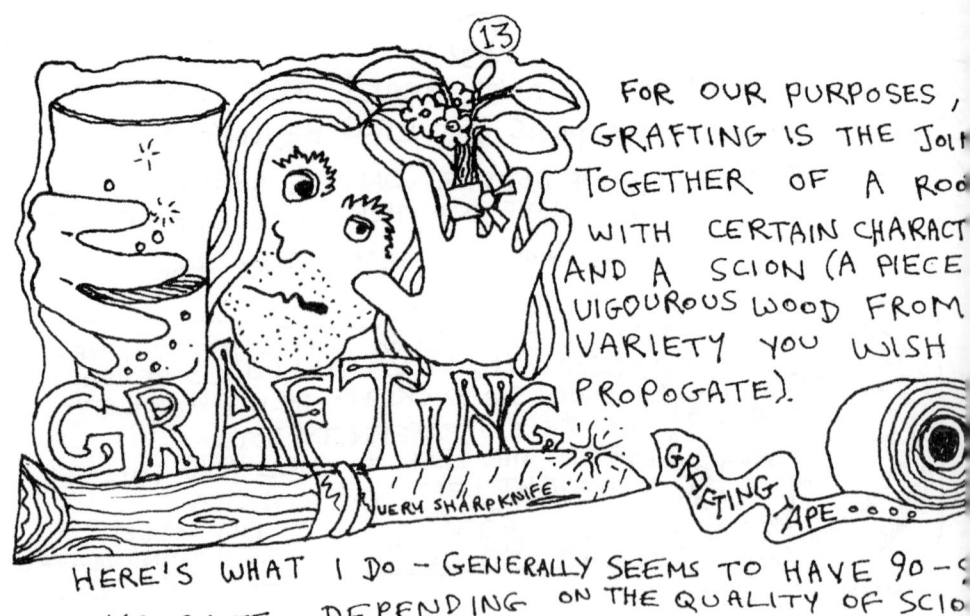

HERE'S WHAT I DO - GENERALLY SEEMS TO HAVE 90-
TAKE RATE, DEPENDING ON THE QUALITY OF SCIO
WOOD. YOU CAN GRAFT ANY TIME THE ROOT STOCKS AND
SCIONS ARE DORMANT - MARCH IS IDEAL IN MOST YEARS.
BE VERY CAREFUL NOT TO CHOP OFF ANY BITS OR
FINGERS OR OTHER STUFF WHILST USING THE VERY
SHARP KNIFE! GRAFT SOMWHERE COOL & DON'T LET
THE CUT ENDS OF SCION & GRAFT DRY OUT
BEFORE JOINING THEM!

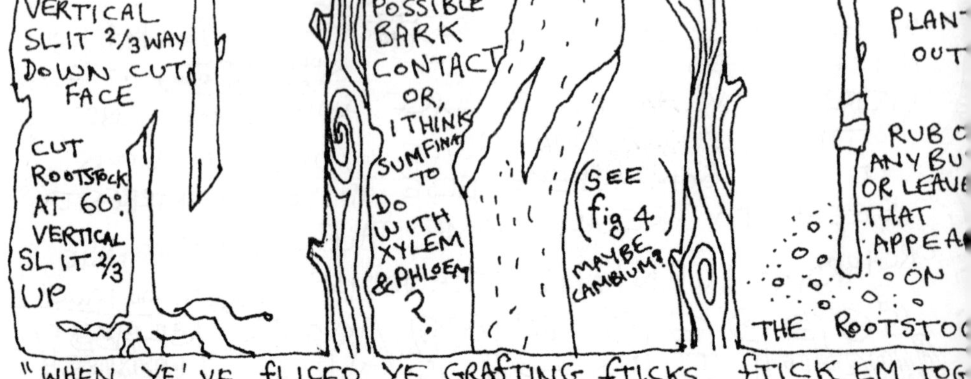
fig 1
CUT BOTTOM OF SCION AT 60° CUT VERTICAL SLIT 2/3 WAY DOWN CUT FACE
CUT ROOTSTOCK AT 60°. VERTICAL SLIT 2/3 UP
PUSH HALVES TOGETHER ENSURING MAXIMUM POSSIBLE BARK CONTACT OR, I THINK SUMFINK TO DO WITH XYLEM & PHLOEM ?
(SEE fig 4)
MAYBE CAMBIUM?
WRAP JOIN WITH SE AMALGAMATING POLYTHENE TAPE
PLAN OUT
RUB ANY BU OR LEAVE THAT APPEA ON THE ROOTSTO

"WHEN YE'VE ſLICED YE GRAFTING ſTICKS, ſTICK EM TOG
A HOLDIN ON TER BITS "o" DEAD ſTICKS." John He

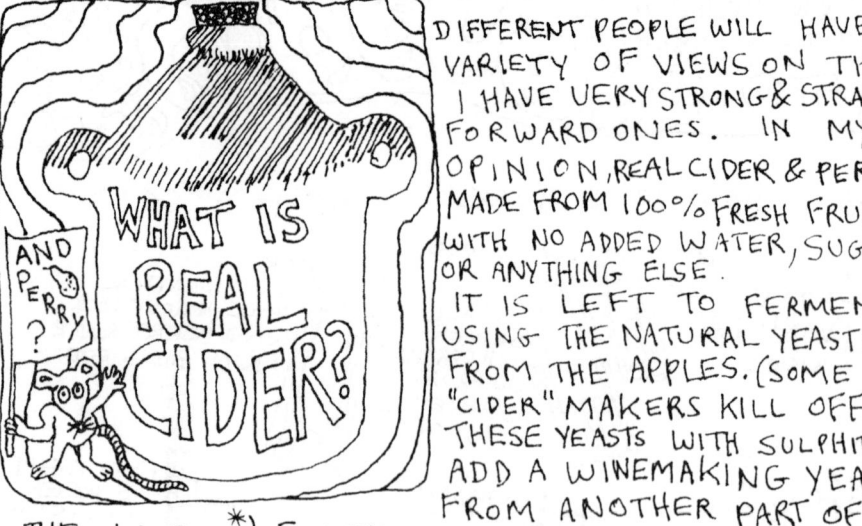

Different people will have a variety of views on this. I have very strong & straightforward ones. In my opinion, real cider & perry is made from 100% fresh fruit with no added water, sugar, or anything else.

It is left to ferment using the natural yeasts from the apples. (Some "cider" makers kill off these yeasts with sulphite & add a winemaking yeast from another part of the world*) Further, I would argue that a strict definition of "real cider" would mean live cider — i.e. cider that has not been sulphited or pasteurized to kill the yeast.

Industrial cider is often less than 20% juice (generally from concentrate), among the rest of the ingredients may be: water, glucose-fructose syrup, saccharin, malic acid, food colouring, carbon dioxide. It is really a "made wine," and should not be confused with real cider

It would be a very good thing if cider makers were compelled to list their ingredients — there are many people who would change their drinking habits if they realized what they were buying was a chemical confection containing less than 30% apple juice.

BRIMSTONE & TREACLE

A NOTE ON THE USE OF SULPHITES & ADDED YEASTS

*I HAVE YET TO TRY A CIDER MADE IN THIS WAY THAT DOESN'T TASTE DEAD AND SLIGHTLY POISONOUS. THERE ARE A NUMBER OF 'CIDER GURUS' OUT THERE PROMOTING WHAT IS ESSENTIALLY A FORM OF MINIATURIZED INDUSTRIAL CIDER MAKING (ALBEIT USING FRESH JUICE RATHER THAN CONCENTRATE). IN MY OPINION, CIDER MADE IN THIS WAY IS A PALE IMITATION OF THE REAL THING. ANOTHER EXAMPLE OF AUTHENTICITY AND NATURAL RICHNESS SACRIFICED FOR THE SAKE OF "CONSISTENCY," PREDICTABILITY AND HOMOGENEITY. YOU CANNOT EXPECT TO EQUAL THE HEARTY COMPLEXITY OF A TRUE CIDER, MADE WITH THE 40 OR 50 INDIGENOUS YEASTS THAT COME IN ON THE APPLES IF YOU KILL THESE OFF AND REPLACE THEM WITH A SINGLE FOREIGN YEAST. FURTHERMORE, YEASTS ARE RESILIENT LITTLE CRITTERS AND I WOULDN'T, PERSONALLY, CHOOSE TO INGEST ANYTHING THAT WOULD KILL THEM.

& SUGARS

THERE IS NO NEED TO ADD SUGAR TO CIDER— UNLESS YOU WANT A BREW OVER 7-8% WHICH TASTES OF ADDED SUGAR.

PERRY

Perry is an alcoholic drink made from pears in much the same way that cider is made from apples. Perry pears tend to be harder and more tannic tasting than dessert pears, and traditional perry tends to be a bit stronger than cider. Perry pear trees can be huge and live for a very long time. I was given some scion wood this winter from some perry pear trees planted at the time of the Napoleonic wars.

Pider

Because pears can grow into very big trees, you may come across large amounts of "Conference" or other dessert pears that the owners are happy to let you have. I would suggest you try making pider with these, as it's really rather good. I think the ideal recipe is something like 70% pear to 30% apple (ideally a bittersweet apple, but any would work.)

A marriage made in heaven

"TUMPING"

"TUMPING" APPLES IS REALLY JUST STORING THEM IN A PILE. IF YOU PASS THROUGH HEREFORDSHIRE IN THE AUTUMN YOU WILL SEE THOUSANDS OF TONS OF APPLES IN CONCRETE LINED TUMP YARDS WAITING FOR PICK UP OR DELIVERY.

APPLES ACTUALLY KEEP PRETTY WELL WHEN PACKED TOGETHER TIGHT, DUE TO THE CO_2 THEY BATHE THEMSELVES IN AS THEY RESPIRE. CIDER QUALITY CAN BE IMPROVED BY ALLOWING GREENISH OR UNRIPE FRUIT TO SIT FOR A WEEK OR MORE BEFORE PRESSING. OBVIOUSLY, ONE SHOULD KEEP AN EYE OUT FOR ROT AND THROW OUT ANY ROTTEN APPLES AS USUAL.

PEARS HARDLY KEEP AT ALL WHEN RIPE, RIPEN VERY QUICKLY AND UNPREDICTABLY, AND YOU'RE BEST PRESSING THEM AS SOON AS POSSIBLE! AVOID THE HEINOUS MISTAKE OF MIXING APPLES AND PEARS IN A DUMPY BAG UNLESS YOU'RE DEFINITELY GOING TO PRESS THEM ON THE SAME DAY, OTHERWISE YOU'LL BE PICKING/ WASHING ROTTEN PEARS OUT OF PERFECTLY GOOD APPLES ALL DAY.

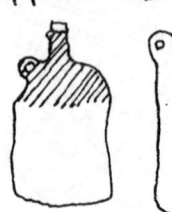  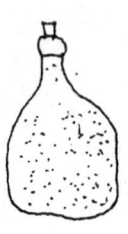   

Some Notes on Barrels

Wooden barrels are very convenient to use in the making of cider. They can be bought wet from distillers, or from sellers in cider areas, from 35 imperial gallon upwards. I use 100 gall bourbon barrels and 110 gall. sherry butts, both of which have also had lives at a scotch whisky distillery. There will be a slight oaky & spirit taste on the cider, which some don't like. I love it. Rum & brandy casks are also available wet, which will again give a different taste to your cider or perry. Part of the convenience comes from the fact that, being soaked in spirit on the inside, all the casks will already be sterile. Casks can either be filled upright, or chocked on their sides. There is generally a bung in the side of the barrel & one in the top, either of which can be used to fill up the barrel. To tap the barrel, drill a hole, the same diameter as the stem of your tap, nearly all the way through the centre of a stave ('til a drip comes), remove drill and bash in tap with a mallet.

TOP TIP: MAKE SURE TAP IS GOOD & DOESN'T LEAK FIRST!

Problems with barrels: outside of staves may dry out in hot weather causing small leaks where they meet — the solution is to dampen the staves with clean water. Barrels seem less prone to this if kept on their sides rather than on their ends.

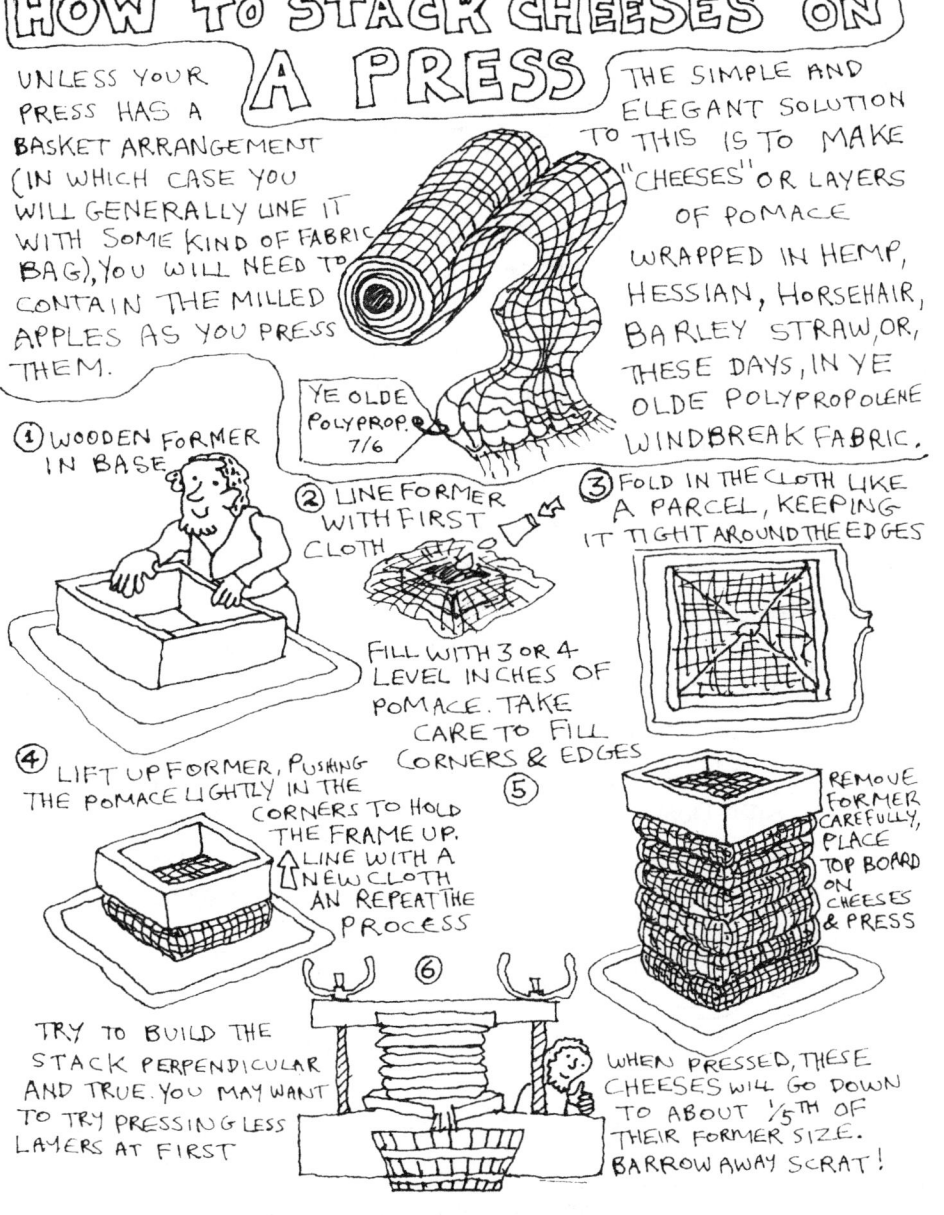

Notes on Equipment

IN THE 18th & 19th, THE CONDITION "CIDER COLIC" WAS WELL KNOWN IN THE WEST COUNTRY. CIDER COLIC IS, IN FACT, ACUTE LEAD POISONING CAUSED BY THE PRACTICE OF LINING CIDER PRESS JUICE TRAYS WITH LEAD. SO LET THAT BE A LESSON TO YOU CAMPERS!!*?! CIDER AND METAL DON'T MIX. NO NICKEL, COPPER, ALUMINIUM, STEEL OR IRON. THE ONLY METAL YOU SHOULD LET NEAR YOUR APPLES IS STAINLESS STEEL, AND I TRY TO KEEP EVEN THAT TO A MINIMUM. A CERTAIN SOMERSET CIDER MAKER, WHO MAKES POSSIBLY THE BEST CIDER IN THE WORLD, GAVE ME 50 PINTS OF AMBROSIAL NECTAR RECENTLY. I DRANK A COUPLE OF PINTS WITH A FRIEND, AND CHEERILY LOOKED FORWARD TO CONSUMING THE REST AT MY LEISURE. ALACK! THE NEXT MORNING THE CIDERTASTIC TIPPLE HAD TURNED INTO A DARKLY GREENY BLACK UNSAVOURY TONIC DUE TO THE OXIDATION OF IRON SALTS PICKED UP FROM THE SCRATTER & PRESS. THE ILLUSTRIOUS CIDER MAKER HAS NOW REPLACED ALL HIS STEEL AND IRON PARTS.

Cleanliness

THE MAIN CONTAMINATION RISK IN CIDER MAKING IS THAT YOU WILL END UP WITH HUNDREDS OF GALLONS OF VINEGAR. PART OF THE DEFENCE AGAINST THIS IS TO WATCH OUT FOR THE VINEGAR FLY — AVOID LEAVING ANY KIT AROUND COVERED IN JUICE, KEEP ALL CONTAINER LIDS ON, COVER ALL OPEN RECEPTACLES WITH NETTING. KEEP KIT CLEAN & STERILIZE WITH MILTON OR SIMILAR DISINFECTANT

Vinegar Fly — This chap is a problem because he carries acetobacter which turns ETHANOL into ACETIC ACID

ALL KIT SHOULD BE THOROUGHLY FLUSHED, SCRUBBED & WASHED AT THE END OF THE DAY. ANY KIT THAT GETS ACCIDENTALLY DROPPED ON THE FLOOR SHOULD BE IMMEDIATELY WASHED BEFORE USE AGAIN.

ALL HOSES SHOULD BE PUMPED THROUGH WITH DISINFECTANT SOLUTION. FERMENTATION VESSELS SHOULD BE STERILISED BEFORE USE ("WET" SPIRIT CASKS ARE ALREADY STERILE)

SMALL SCALE PRODUCTION
A COUPLE OF PLASTIC FERMENTING BINS & SOME DEMI-JOHNS OR BOTTLES WILL SUFFICE

CIDER IS GENERALLY VERY ACIDIC AND QUITE ALCOHOLIC. THIS IS WHY THERE HAS NOT BEEN ONE CASE OF MICROBIAL GASTRIC INFECTION FROM REAL CIDER. HOORAH!
THO' A SURFEIT CAN STILL AFFECT SOME ADVERSELY — WRONG BOW

LARGER SCALE PRODUCTION
WET SPIRIT BARRELS
1000L POLY IN CAGE
225L PLASTIC OLIVE BARRELS

HAVIN SAID ALL THAT, IT SEEMS LIKE IT'S ACTUALLY QUITE HARD TO MAKE A BAD BATCH. IF YOU DO, CIDER VINEGAR IS TAX FREE & DUTY FREE, AND IS WORTH AS MUCH OR MORE THAN CIDER! OR YOU COULD GROW GHERKINS & START A PICKLE FACTORY.

OTHER THINGS TO MENTION ARE JUST COMMON SENSE, REALLY. KEEP YER HANDS CLEAN AND AVOID CONTAMINATING THE APPLE/POMACE/CIDER WITH DIRT OR FOREIGN MATERIAL.

FASCINATING FERMENTATION FACT
MOST OF THE ALCOHOL IN CIDER IS FORMED IN THE FIRST WEEK OR TWO AFTER PRESSING. MOST PROCESSES SHUT DOWN IN THE COLD OF WINTER. AS THE SPRING WARMS UP, IN A SERIES OF SECONDARY FERMENTATIONS, THE LACTIC ACID IS TURNED INTO LACTALOSE — A SUGAR THAT IS NOT BROKEN DOWN BY YEAST. THIS MELLOW'S THE CIDER'S TASTE

Ageing, Keeping & Cider

Now you have your juice in barrels, how to keep it, and how long to keep it for?

Traditionally, it is said that last year's cider is not ready to drink 'til you hear the first cuckoo of spring. Readiness will depend on factors such as the size of the container and the temperature where it is stored. Essentially, it's ready when it tastes good! This could be after 2 months or 2 years. Personally, I prefer a bone dry clear cider that's been kept for at least a year, but some like a cloudier, sweeter, young cider.

To Rack off or Not to Rack off

Some cider makers like to rack off (i.e. syphon or pump off into new containers) their cider after the primary fermentation has finished – usually 2 or 3 weeks. This removes the cider from the lees – the yeast and solids that have been precipitated out of the cider. This may slow down or stop fermentation giving a clear, sweet cider – particularly if racked off repeatedly, removing most of the yeast – A more sophisticated form of this, that would merit a book in its own right, is the practice of keeving used in Brittany and Normandy

I GENERALLY DON'T BOTHER RACKING OFF CIDER AFTER THE PRIMARY FERMENTATION. SOME PEOPLE ARGUE THAT CIDER CAN PICK UP "OFF" FLAVOURS FROM SITTING ON THE LEES. I HAVEN'T FOUND THIS TO BE THE CASE — I HAVE CIDER THAT HAS BEEN SITTING ON THE LEES FOR TWO YEARS THAT STILL TASTES GREAT. ON THE OTHER HAND, RACKING OFF A FEW DAYS OR WEEKS BEFORE DRINKING CAN CERTAINLY IMPROVE AND GIVE A MORE ROUNDED TASTE TO CIDER. A SMALL AMOUNT OF OXIDATION ENRICHES FLAVOUR — TOO MUCH CAN RUIN IT.

A DISADVANTAGE OF HABITUALLY RACKING OFF CIDER IS THAT AT EVERY POINT YOU ARE RISKING CONTAMINATION BY ACETOBACTER ET AL — TAKE CARE!

IF YOU ARE GOING TO SELL OR DRINK YOUR CIDER AWAY FROM YOUR HOUSE, YOU'LL NEED TO PUT YOUR CIDER IN SMALLER CONTAINERS. PRESSURE BARRELS ARE RE-USABLE. BAG IN BOX (LIKE A GIANT WINE BOX) ARE RELATIVELY CHEAP, HYGENIC, AND CAN BE ATTACHED TO PUB BEER LINES. GLASS BOTTLES ARE POPULAR, BUT EXPENSIVE AND WASTEFUL UNLESS RE-USED. PLASTIC "MILK" BOTTLES ARE CHEAP, BUT WON'T KEEP SO LONG & LOOK CHEAP!

40 pint pressure barrel

20/10/3 litre Bag in Box

DON'T BE ALARMED IF YOU OPEN A VAT OF CIDER AND IT HAS A WHITISH BLOOM, BROWN YEASTY CAP OR OTHER FUNKY LOOKING STUFF FLOATING ON IT. IT IS, AFTER ALL, A LIVING THING. WHITE FLOATERS ARE PECTINATED YEAST & ARE A SIGN OF QUALITY.

YOU CAN EITHER KEG/BOTTLE OFF A WHOLE BARREL IN ONE GO (PARTICULARLY RECOMMENDED WITH PLASTIC BARREL) OR RUN OFF CIDER THROUGH A TAP ON DEMAND. I'VE HAD BARRELS TAPPED FOR MORE THAN 6 MONTHS, AND IT'S ALWAYS BEEN GOOD TO THE LAST DROP. WITH A SPIRIT CASK, THE CIDER GETS PROGRESSIVELY MORE SPIRITOUS AS SPIRIT RICH CIDER RUNS OUT OF THE INSIDE OF THE WOOD INTO THE REMAINING CIDER.

CROWN CORKED GLASS BOTTLE

2 Litre "milk" bottles

Selling Your Cider & Perry (29)

IF YOU'RE MAKING DELICIOUS CIDER, WHY NOT SELL IT? SELLING ANYTHING YOU MAKE WILL INTRODUCE A WHOLE NEW EXCITING ASPECT (OF DEALING WITH THE ORGANS OF THE STATE) INTO YOUR LIFE. EXPLICITLY: |H.M. CUSTOMS| WILL WANT TO KNOW HOW MUCH CIDER YOU ARE MAKING AND SELLING. AT PRESENT (2011) THERE IS A DUTY EXEMPTION FOR SMALL CIDER MAKERS WHICH MEANS THAT IF YOU SELL 15½ HUNDRED GALLONS OR LESS PER YEAR (70 "HECTOLITRES" OR 7,000 LITRES), THERE IS NO DUTY TO PAY. OBVIOUSLY, H.M. CUSTOMS ARE VERY KEEN THAT YOU KEEP ACCURATE AND UP TO DATE RECORDS OF ALL SALES, AND MAY CALL ON YOU AT ANY TIME TO CHECK UP.

|TRADING STANDARDS| WILL BE INTERESTED IN YOUR PACKAGING & LABELLING & WHETHER IT'S UP TO REGULATIONS. THEY MAY TEST THE ALCOHOL CONTENT TO SEE IF IT'S AS CLAIMED (H.M. CUSTOMS MAY ALSO BE INTERESTED IN THIS!)

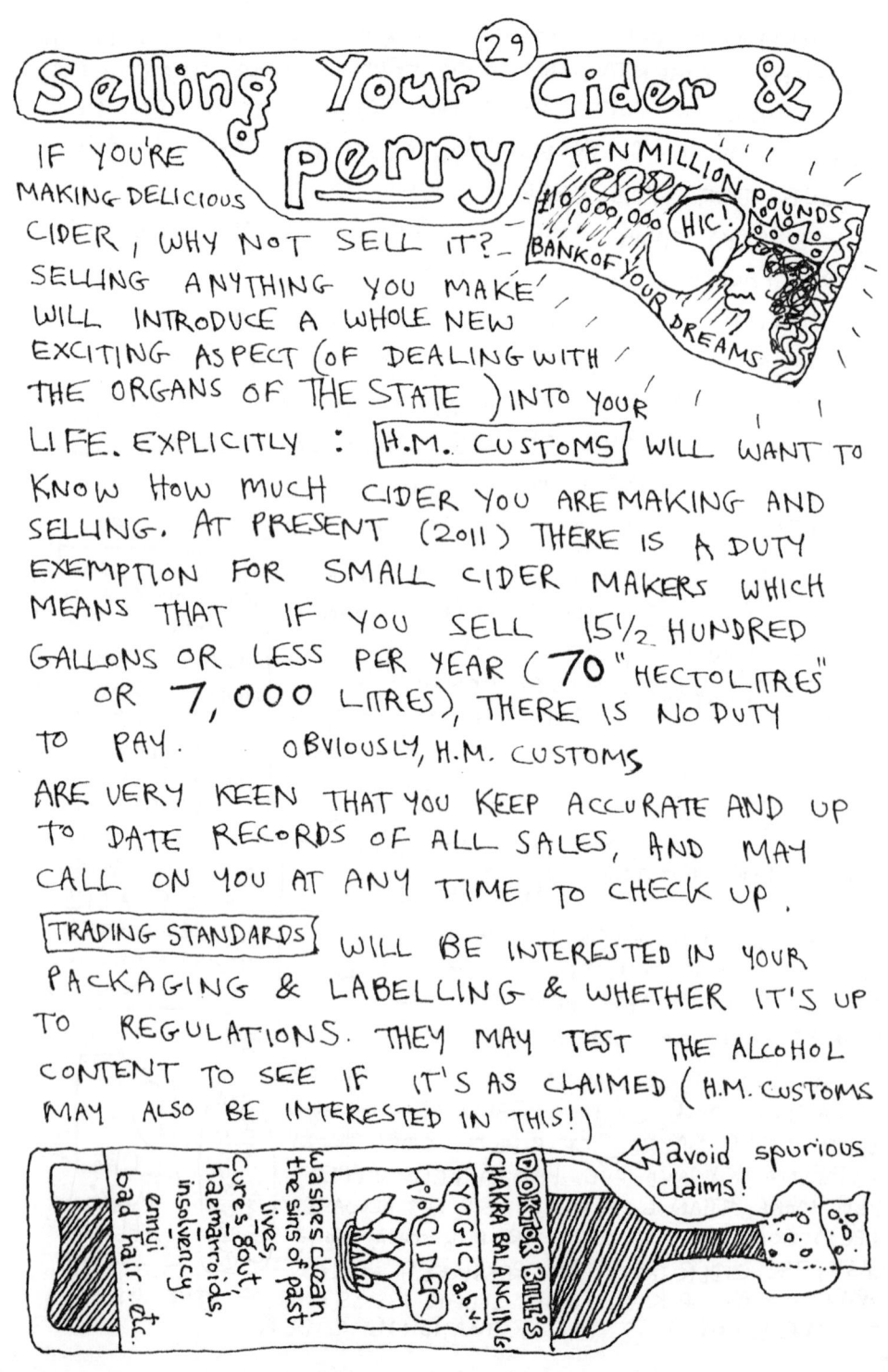

avoid spurious claims!

ENVIRONMENTAL HEALTH "DEEPJOY" AS SIR STANLEY UNWIN ONCE SAID. ENVIRONMENTAL HEALTH WILL HAVE A LOT OF IDEAS ABOUT HOW YOU SHOULD BE DOING EVERY THING. THESE WILL VARY SOMEWHAT ACCORDING TO THE LOCAL AUTHORITY AREA YOU ARE IN. IT IS VERY MUCH WORTH CONTACTING THEM BEFORE YOU PUT ANY WORKS IN PLACE FOR YOUR PROJECT SO THAT YOU DON'T MAKE MISTAKES THAT COULD BE COSTLY & TIME CONSUMING TO RECTIFY LATER. YOU WILL BE EXPECTED TO HAVE CONDUCTED A RISK ANALYSIS, AUDITED YOUR PRODUCTION LINE HYGEINE, ETC.

IF YOU HAVE A PRIVATE WATER SUPPLY, YOU WILL HAVE TO HAVE IT TESTED BY THE COUNCIL EVERY YEAR.

NEEDLESS TO SAY, ALL THE PEOPLE I HAVE ENCOUNTERED IN THE ABOVE SERVICES HAVE BEEN A DELIGHT AND A JOY TO DEAL WITH, AND I LOOK FORWARD TO WORKING WITH ALL OF THEM TOWARDS A BRIGHTER, SAFER FUTURE.

SELLIN CIDER 'N' DA LAW — IF YOU WISH TO RETAIL ALCOHOL, YOU WILL NEED A PERSONAL LICENSE. TO GET THIS YOU WILL HAVE TO ENROL ON A ONE DAY PERSONAL LICENSE COURSE, WHERE THEY'LL TELL YOU EVERYTHING ELSE YOU'LL NEED. SO I WON'T.
 NOW YOU'VE GOT YOUR PERSONAL LICENSE, YOU PROBABLY THINK THAT YOU'RE THERE! NOT QUITE. YOU STILL HAVE TO OPERATE FROM A PREMISES OR LOCATION THAT HAS EITHER A TEMPORARY OR PERMANENT LICENSE. ACH, THEY'LL TELL YOU ON THE COURSE!

I HOPE YOU ENJOYED MY LITTLE BOOK. IT SHOULD GET YOU GOING — YOU WILL EVENTUALLY FIND THE BEST WAYS OF DOING THINGS FOR YOUR SET UP BY EXPERIENCE (FOR WHICH THERE IS NO SUBSTITUTE!)

¡HAPPY JUICING, ORCHARDISTAS!

FORTHCOMING TITLES FROM

— WELSH MOUNTAIN BOOKS —

"HOW TO MAKE STAINED GLASS & LEADED LIGHTS."

FUTURE PROJECTS MAY INCLUDE:

"BILL'S ELABORATE POMONA"
—

"SQUIRREL BILTONG — HOW TO DRY VERMIN FOR PLEASURE & PROFIT"
—

"GOING CRITICAL — BACK GARDEN THERMONUCLEAR REACTOR BUILDING FOR THE AMATEUR ENTHUSIAST."

NATURAL CRAFT CIDER MADE FROM 100% APPLES
WITH NO ADDED SUGARS, YEASTS OR SULPHITES
AVAILABLE IN 20lt BAG-IN-BOXES FROM
PSYDWR.COM

7%a.b.v.

NatSol
COMPOST TOILET SPECIALISTS

Call Ellen,
Chris or
Andy
01686 412653
natsol.co.uk

No water, no power, no smell, no worries

Over 200 installations:
Allotments,
orchards,
campsites,
smallholdings,
churches,
gardens...

Full wheelchair access
Steel, Timber or
DIY building.
Urine separating design
handles high use

The Clean Energy Centre
(Newtown)

Mid-Wales Electric Bike Specialists

The most fun (and greenest) way to travel

Take a ride on one of our fleet of demonstration bikes.

We also supply and fit kits to convert your own bike

The Clean Energy Centre
Gas Street, Newtown
SY16 2AD
07722 722863

www.cleanenergycentre.org.uk

If you would like to place an advert in the next edition of this book, call Bill on: 0779 0071729